CAN YOU FIND:

- ☐ FROG EGGS THAT LOOK LIKE A CLUSTER OF SEE-THROUGH GRAPES MADE OF JELLY?
- ☐ NEWLY LAID WOOD FROG EGGS, WHICH HAVEN'T EXPANDED WITH WATER YET?
- ☐ SPRING PEEPER EGGS, WHICH LOOK LIKE TINY SINGLE PEARLS CLINGING TO UNDERWATER PLANTS?
- ☐ TOAD EGGS, IN A LONG RIBBON OR TUBE?
- ☐ SPOTTED SALAMANDER EGGS—A LARGE CLUSTER ENCASED IN AN EXTRA-THICK OUTER LAYER OF JELLY?
- ☐ JEFFERSON OR BLUE-SPOTTED SALAMANDER EGGS—A LOOSER, LONGER MASS WITH FEWER EGGS THAN SPOTTED SALAMANDER CLUSTERS?
- ☐ SPOTTED SALAMANDER SPERMATOPHORES, WHICH LOOK LIKE BREADCRUMBS?

For Jasper, Ulysses, and Greg:
Thank you for all the magical adventures in the rain.

Library of Congress Cataloging-in-Publication Data available.

ISBN 978-1-7972-1456-6

Manufactured in China.

Design by Eugenia Yoh.
Typeset in Ernestine Pro and Palmer Lake Print.
The illustrations in this book were rendered digitally, informed by woodcut printmaking techniques.

10 9 8 7 6 5 4 3

Chronicle Books LLC
680 Second Street
San Francisco, California 94107

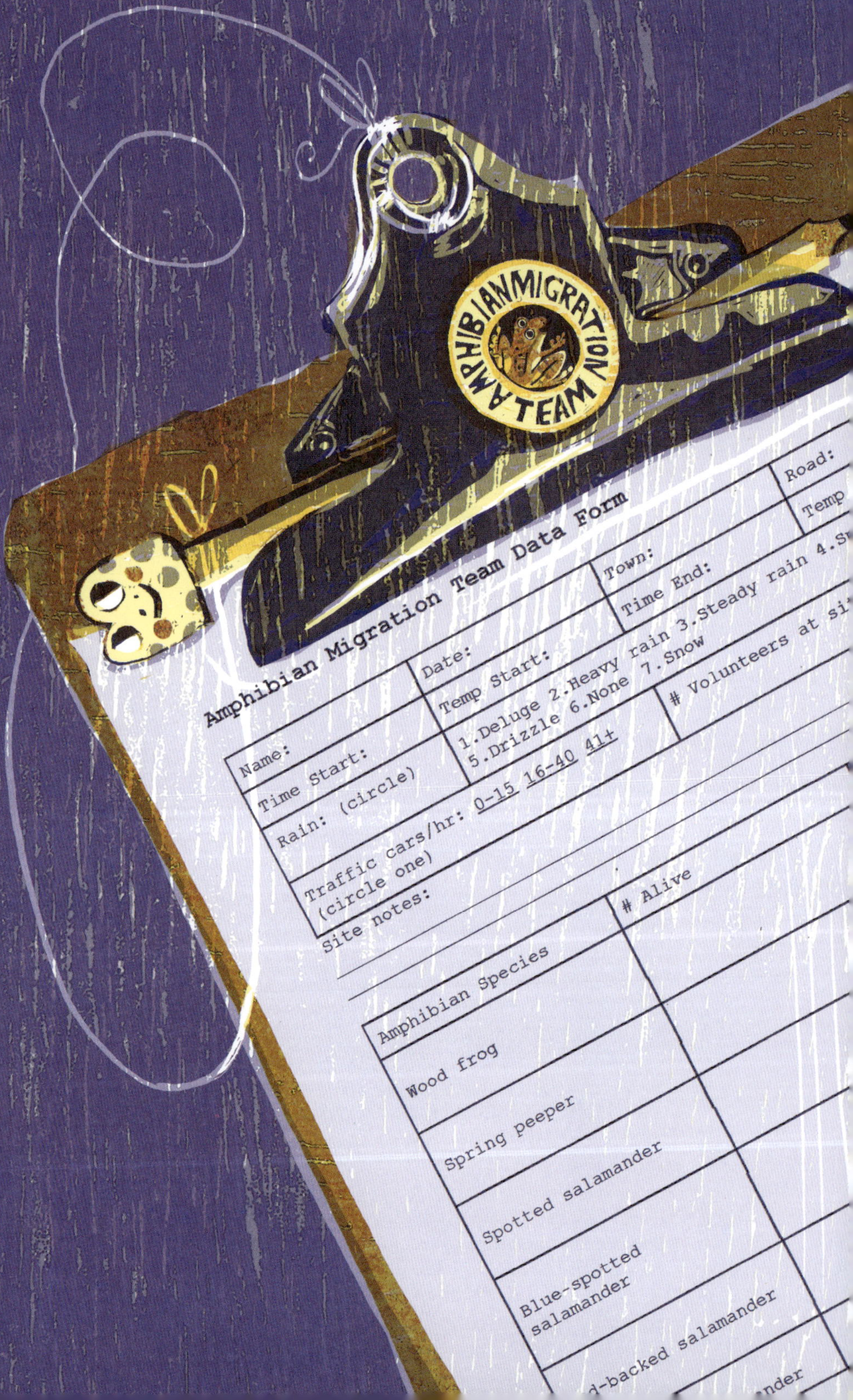

SAFE CROSSING

KARI PERCIVAL

CHRONICLE BOOKS
SAN FRANCISCO

How do they know it's time?

Each spring, during the night
of the first soaking rain after the thaw,
the amphibians know.

And guess what?

The Amphibian Migration
Team knows it's time, too!

Time to pull reflector vests and raincoats on over our pajamas. Time to find our headlamps and bring our flashlights down to where the spring peepers are singing so loud, it makes your spine tingle—

down to the crossing place:
a wet road on a wet night in spring.

We are the crossing crew, this rainy night.

We got the town to put up a detour,
but some people still drive this way
because they live over here.
So we still have to watch out for cars.

During breaks in the traffic,
we help our tiny friends cross the street.

The wood frogs, spring
peepers, salamanders—
spotted, blue-spotted,
and Jefferson—
and other amphibians

are crossing the road
tonight:

a wet road on
a wet night in spring.

We rinse our hands in cold water first.
Then we look for tiny creatures on the wet pavement.

"But how do they know tonight is the night?" asks Tallulah.

"The spring equinox wakes them," I say.

"But *how*? Do they taste the warm rain? Do they smell the skunk cabbage blooming?" asks Tallulah.

We see salamanders and frogs on the road, pick them up gently, and give them a ride on wet leaves to bring them in the direction they want to go. Some are going to the pools; some are already crossing back up to the woods again.

"Do they hear the other frogs singing and think, *Tonight is the night! It's time to go down to the pool to lay our jelly eggs safely in the water! Let's go!*? But then how do the other frogs know to start singing?"

UPLAND WOODS

I don't know.

I tell her how they travel
down from the woods,
out of the holes
under big tree roots,
out from under
old logs.

They are going down to the
vernal pools beyond the road.

“And how do they know *where* to go?” asks Tallulah. “Dad, how do the amphibians know to cross the street *here*?”

Dad says, “I bet this is the path they have always taken from the woods down to the vernal pools and back. Long before people paved this road for cars. Long before cars and roads.”

“They must be going back to where *they* first hatched out of *their* eggs, where *their* mothers laid jelly eggs in the water,” I think out loud.

“Car coming!” calls Mom. Everyone scatters to the edges of the road.

"This *road* is in the way!" says Tallulah.

"Why did people have to build a road here? These creatures are so tiny. The tires are so big. The cars go so fast. And what about times they cross when we aren't here to help them?"

We count them: how many cross,
how many don't make it in time.

The spring peepers are singing so loud,
it makes my spine tingle.

"Dad, I wish there were a way they could *all* make it across," I say.

"Me, too!" says Tallulah.

Dad says he wishes there were a way, too.

We write the numbers on the clipboard.

"Dad, can we *make* a way?" I ask.

He says, "What's your idea?"

I say, "They need their own safe crossing. Their own bridge or their own tunnel."

Dad says, "I think you are on to something."

A WILDLIFE BIOLOGIST MAPS WHERE THE MOST AMPHIBIANS GO.

THE CONSERVATION COMMISSION GETS PERMITS AND HIRES CONTRACTORS TO BUILD THE TUNNELS.

Dad looks into it.
It turns out that wildlife tunnels are already a real thing!

There is a lot to do, and we need a lot of help.

THE DEPARTMENT OF FISH AND WILDLIFE HELPS ORGANIZE OUR DATA AND TELLS US HOW TO APPLY FOR A WILDLIFE GRANT.

CONTRACTORS PRICE OUT ALL THE PARTS OF THE DESIGN.

THE DEPARTMENT OF TRANSPORTATION TELLS US HOW TO APPLY FOR A ROAD GRANT AND MAKES SURE OUR PROJECT WILL WORK.

A HERPETOLOGIST AND A ROADWAY ENGINEER DESIGN THE TUNNELS TO BE AMPHIBIAN FRIENDLY.

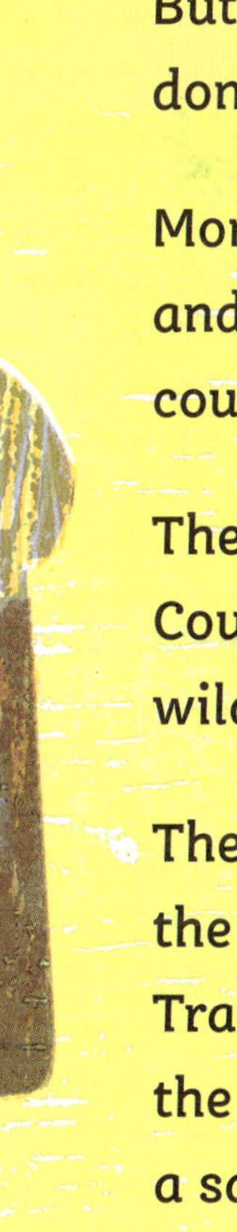

But we have already done part of the work.

Mom says all those frogs and salamanders we counted add up:

They could show the Town Council how important a wildlife tunnel would be.

They could show the Department of Transportation why the animals need a safe crossing.

REPORTERS ASK QUESTIONS AND SHARE OUR IDEA WITH EVERYONE!

A SURVEYOR MEASURES THE HILL.

Our project is first up at the hearing.

My belly feels like it's full of jumping frogs!
I read what I wrote on index cards.
My voice shakes a little,
but the council is so quiet,
listening.

“Guess what? We are getting our wish! They voted yes! And we got the grant!”

But we have only won half the money we need to build the tunnels.

There are no tunnels yet, but spring comes anyway.

The Amphibian Migration Team is bigger this year. More people want to help. I show them how to wash their hands in the rainwater first.

Everyone is asking: “Where can we get the other half of the money?”

Everyone thinks of creative ideas to raise money for the wildlife crossing:

The scout troop collects cans.

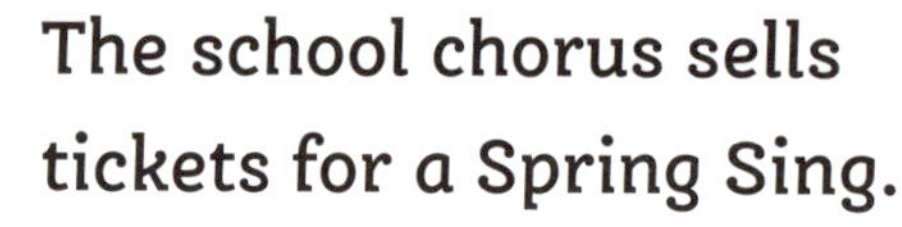

The school chorus sells tickets for a Spring Sing.

The senior center holds a bake sale.

The Rotary Club runs
a rummage sale.

Our art class auctions our paintings.

The Conservation Commission asks people
to walk, bike, or carpool to work or school for
a week and donate the price of a tank of gas.

The Amphibian Migration Team sells
"Make Way for Wildlife" buttons
and bumper stickers.

All that money adds up.

And guess what?
We raise enough!

Late in the summer, we show the construction workers where to slice channels in the pavement. Big machines dig trenches and lower concrete tunnels into place. Workers shovel sandy soil into the bottom. They install grates on top to let in the rain. They add concrete guide walls so the amphibians will know where to go.

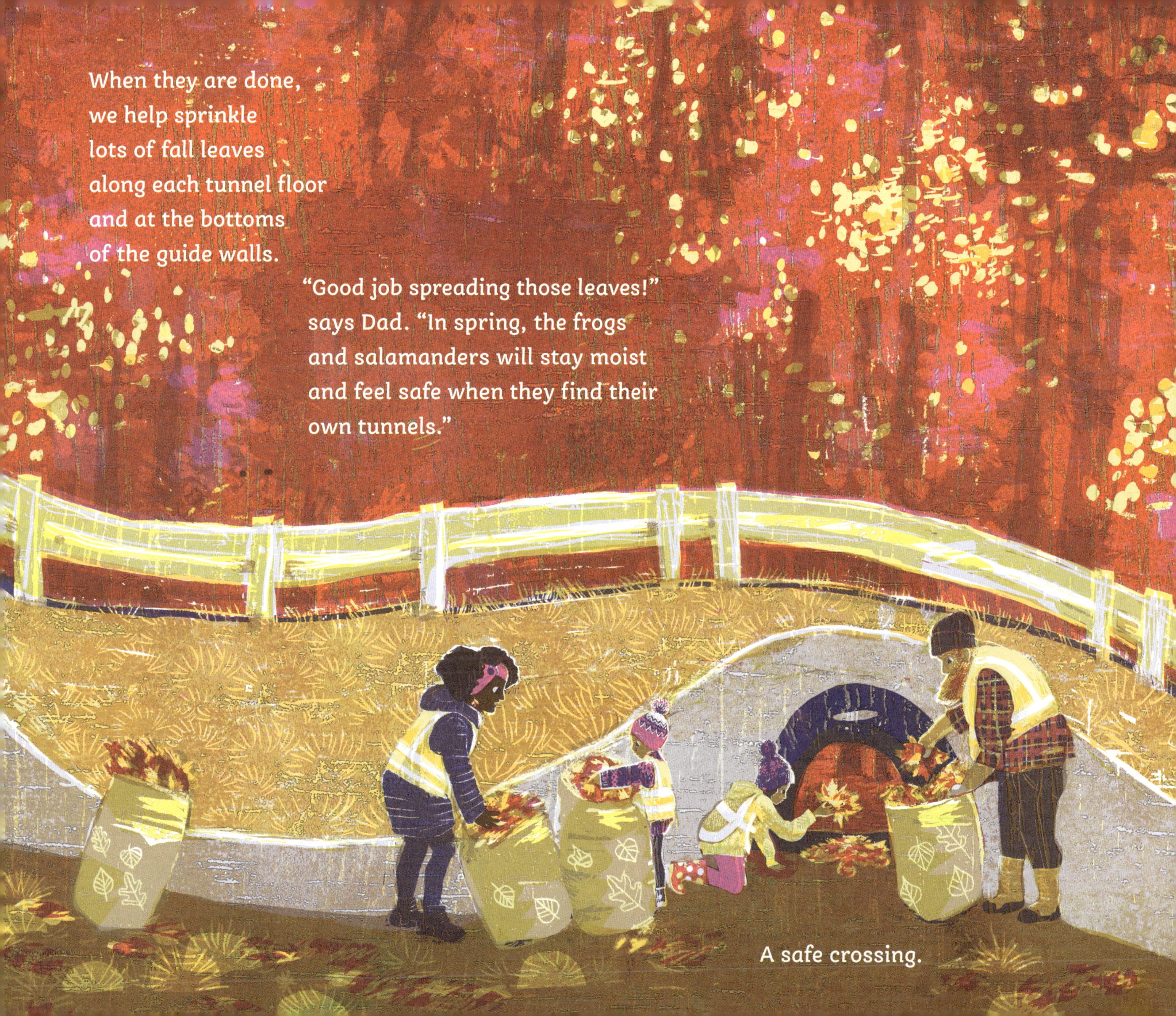

When they are done,
we help sprinkle
lots of fall leaves
along each tunnel floor
and at the bottoms
of the guide walls.

"Good job spreading those leaves!" says Dad. "In spring, the frogs and salamanders will stay moist and feel safe when they find their own tunnels."

A safe crossing.

In autumn, the rains fill
the vernal pools again.

As the forest floor freezes,
the spotted salamanders
creep deeper underground.

In winter, fairy shrimp hatch
under the pools' ice,

while in the uplands, the spring peepers keep just warm enough in the leaves under the snow, and the wood frogs freeze almost solid for a while.

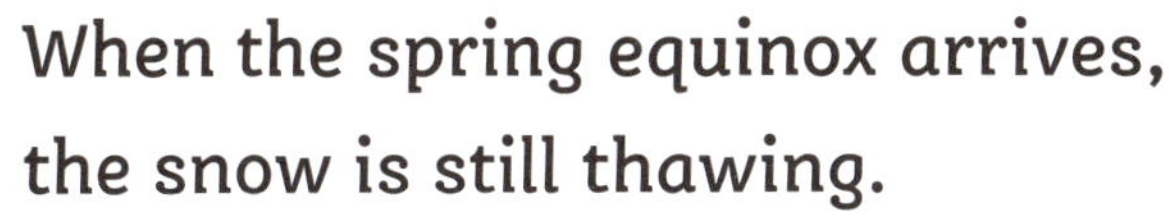

When the spring equinox arrives,
the snow is still thawing.

A heavy spring rain soaks the forest floor.

We bring our flashlights out again into the rain, down to where the spring peepers are singing so loud, it makes your spine tingle.

This time, we count the ones
crossing the road *and*

all the ones that go through the tunnels and come out the other side to freedom, to new families, to safety, at the safe crossing place:

a wet road on a wet night in spring.

AMPHIBIAN LIFE CYCLES

As seasons change and amphibians grow through different life stages, they sometimes need places that are dry, and sometimes wet!

As adults, some amphibian species live on land. They hide in wet leaves, under logs, or up in trees, or they burrow under tree roots and hunt for food such as worms and bugs.

Many species seek water in which to lay their jelly eggs. Vernal pools are a great choice because they are filled with water in spring and fall, but they dry up in summer. Fish cannot survive in pools that don't hold water all year round. This means the amphibians are safe from fish eating up all their eggs. The same frogs, toads, and salamanders might lay eggs in the same vernal pools for years and years.

After the jelly eggs hatch, many of the amphibian larvae live in water while they grow front and back legs. Many use gills to breathe underwater until they develop lungs to breathe air on land. They eat green algae or tiny water creatures like fairy shrimp and mosquito larvae. Some species of larval salamanders, frogs, and toads can eat hundreds of mosquito larvae in one night!

Most juvenile amphibians, such as young spotted salamanders, crawl out of the water and live underground while they grow. As adults, they return to vernal pools to court and lay their eggs.

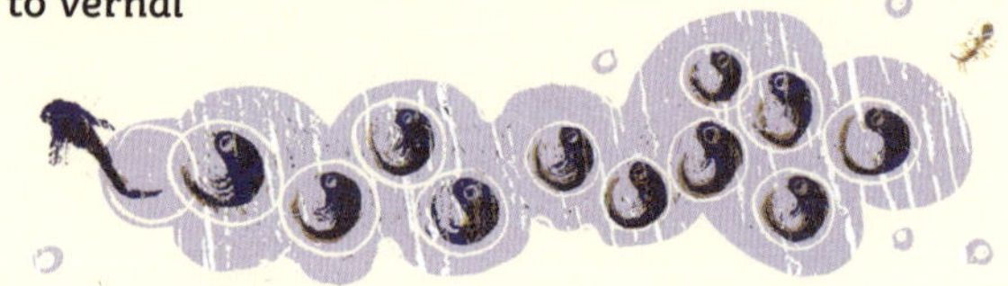

AMPHIBIAN SPECIES FUN FACTS

Spotted salamanders can live to be 20 or 30 years old if they survive to adulthood!

Jefferson salamanders can come out in the snow! Some females clone themselves.

Blue-spotted salamanders can ooze a milky toxin from under their tails to thwart predators.

Wood frogs can freeze almost solid and come back to life. And they quack like ducks.

Spring peepers sing SO loudly for creatures SO tiny—90 decibels!

Gray tree frogs look like lichen and have suction-cup fingers!

American toads are bumpy and lumpy. One toad can eat 1,000 insects and other creatures in one night! They also shed and eat their own skin!

Spadefoot toads spend most of their lives buried underground. Females can lay up to 2,500 eggs at once.

AMPHIBIAN CROSSINGS

Amphibians need our help. Paved roads often split the lowland vernal pools and other wetlands they use in spring and summer from the upland woods they use the rest of the year.

People just like you find creative strategies to help protect amphibians.

WITH YOUR FAMILY:

- Stay home and don't drive on rainy spring nights! Spread the word and encourage friends and neighbors not to drive on those nights, either.
- Observe and help amphibians cross the road on migration nights.
- Track how many and which species you see crossing.

TEAM UP WITH OTHERS:

- Share your data with a local amphibian conservation project or a state wildlife agency.
- Use your data to coordinate with the local department of public works, police, and other town officials to close roads and set up detours at key crossing sites on rainy spring nights.
- Use your data to make the case for protecting surrounding habitats from development.
- Advocate for installing wildlife tunnels or bridges in the event of scheduled roadway improvements.
- Help apply for grants to fund the planning and installation of wildlife tunnels or bridges.
- Keep observing amphibian crossings!

WILDLIFE CROSSING STRUCTURES

Where is a good place for a wildlife tunnel or bridge?

If there is a species that needs help, such as a threatened or endangered species (whose population is at risk of dying out or needs more space to survive), AND if that species has to travel from one habitat to another to complete its life cycle, AND if there is a busy road in the way, so that getting hit by cars hurts the population's chances for survival, THEN it might be possible to get government funding to build a wildlife tunnel or bridge on that road.

Different types of tunnels or bridges work best for different animals of different sizes. Scientists and engineers work together to plan what kinds of crossing structures to build. Then the scientists monitor the tunnels to see how many animals use them to cross safely.

ROAD SAFETY

When you are helping amphibians cross the road, follow these rules to stay safe:

- Use a reflector vest, bright flashlight, and headlamp, so drivers can see you.
- Only children old enough to understand and follow road safety rules should come.
- An adult-to-child ratio of 1:1 is required. Each child should bring one adult focused exclusively on that one child's safety.

- Stay on the edge of the road while cars pass.
- Never try to stop cars.
- Never run into the street to rescue a creature when a car is coming.

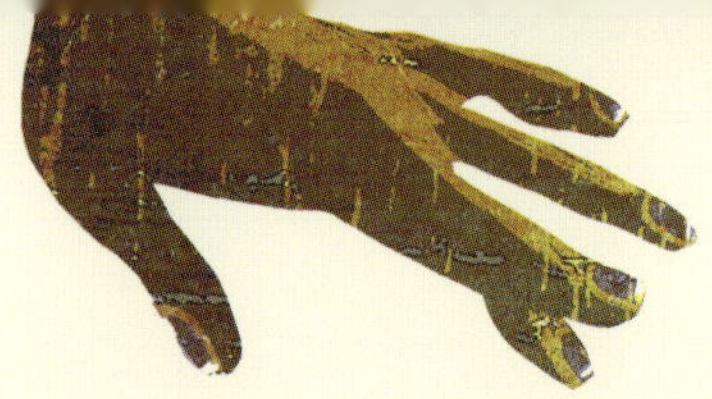

AMPHIBIAN SAFETY

Picking up amphibians stresses them. Only pick them up if they need help crossing the road.

Always wash your hands in cold, clean rainwater before handling amphibians.

Do NOT use hand sanitizer, bug spray, insect repellent, or lotion. Substances like these contain chemicals that can harm the animals.

Carry amphibians on a wet leaf if possible.

Handle amphibians gently, and release them on the ground carefully in the direction they are headed. Do not toss amphibians.

Wash your hands well after handling amphibians, too.

BE A COMMUNITY SCIENTIST

Whether you live in the city or the country, there are community science projects, also called citizen science projects, for people of all ages to join. They allow volunteers to team up with scientists to help all kinds of wildlife.

There are many types of projects you can join, from counting and measuring urban street trees to watching bees and butterflies on flowers to counting birds, turtles, or fish. Each community science project has its own directions about how and when to observe wildlife and how to turn in your data, but all are designed to make it easy for anyone to help. If you do a good job, the data you collect can help show decision makers what changes need to be made to help local wildlife survive.

It's also a great way to meet your neighbors, human and wild, and it feels good to help plants and animals. Who knows where it may lead?

To find a community or citizen science project near you, contact your local environmental education center, natural history museum, zoo, or wildlife conservation organization.

TO GET STARTED, EXPLORE:

- Association of Zoos and Aquariums' list of citizen science projects
- Cornell Lab of Ornithology's list of citizen science projects
- National Audubon Society's list of community science projects
- National Geographic Education's list of citizen science projects
- SciStarter's database of citizen science projects, searchable by topic and location

FOR MORE ON AMPHIBIANS, EXPLORE:

- FrogWatch USA, administered by the Akron Zoo, with local chapters all over the United States
- Harris Center for Conservation Education's list of regional amphibian crossing brigade programs
- Your state or provincial Amphibian Atlas, Amphibian and Reptile Atlas, or Herp Atlas

GLOSSARY

Algae (AL-jee) (pl.): Plantlike living things, often green, often living in water, and often very tiny. Algae grow inside some amphibian eggs, providing oxygen and food for developing larvae.

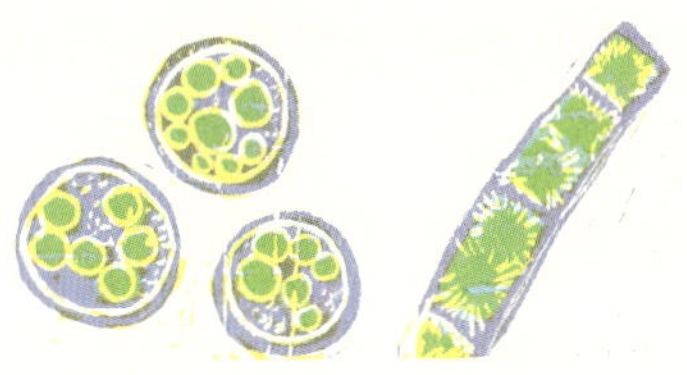

Amphibian (am-FIB-ee-uhn): A moist-skinned and cold-blooded, or ectothermic, animal that requires both land and water to complete its life cycle. Amphibians include frogs, toads, and salamanders.

Community scientist (kuh-MYOO-ni-tee SAHY-uhn-tist): A person who volunteers to make observations or collect data; also called a citizen scientist.

Conservation (kon-ser-VEY-shuhn): The act of protecting and restoring biodiversity of wild species and their habitats.

Data (DEY-tuh or DAT-uh) (pl.): Information collected through observation.

Endangered species (en-DEYN-jerd SPEE-sheez): A species at risk of extinction due to habitat loss or other threats.

Fairy shrimp (FAIR-ee shrimp): A tiny crustacean that lives in vernal pools and swims on its back. This is an important food source for salamander larvae and many other creatures.

Habitat (HAB-i-tat): The natural environment where a plant or animal makes its home.

Larvae (LAHR-vee) (pl.): Animals at a young life stage that go through metamorphosis before becoming adults. Larvae have different body types and habitat needs than adults of the same species. Tadpoles are the larvae of frogs and some toads.

Observation (ob-zur-VEY-shuhn): The use of one of the senses (sight, hearing, touch, smell, or taste) or measurement to gather information.

Pollywog (POL-ee-wog): A frog or toad larva; also called a tadpole.

Population (pop-yuh-LEY-shuhn): The total number of individuals of the same species in an area.

Salamander (SAL-uh-man-der): An amphibian with a long tail, a slender body, and often short front and back legs.

Scientist (SAHY-uhn-tist): Someone who asks questions, makes hypotheses, and tests those hypotheses by gathering data in a systematic way, in order to gain and share knowledge.

Species (SPEE-sheez): A group of living things that can produce offspring (have babies) with others of the same kind.

Spermatophore (spur-MAT-uh-fawr): A packet of sperm left by a male for a female to fertilize her eggs with. Spotted salamander spermatophores look like bread-crumbs at the bottom of a vernal pool.

Survival (ser-VAHY-vuhl): Continuing to live and produce offspring into the future.

Tadpole (TAD-pohl): A frog or toad larva; also called a pollywog.

Vernal pool (VUR-nl POOL): A shallow pond or puddle that is filled by rain or melted snow in spring and fall, with no stream in or out, and that dries up in summer. Because vernal pools dry up, fish cannot live in them and so cannot prey on amphibian eggs laid in them.

Volunteer (vol-uhn-TEER): A person who donates their time or talents to an organization or project.

Wildlife (WAHYLD-lahyf): Animals that are neither domesticated nor human and that live independently and naturally in an area.

CAN YOU FIND:

- [ ] SPRING PEEPER TADPOLES WITH SPOTS, HATCHING OUT OF JELLY EGGS ATTACHED TO AQUATIC PLANTS?
- [ ] A SPRING PEEPER THAT IS JUST DEVELOPING FRONT AND BACK LEGS BUT STILL HAS A TAIL?
- [ ] FIVE-DAY-OLD WOOD FROG TADPOLES, WHICH LOOK LIKE FUNNY LITTLE OWLS WITH LONG TAILS?
- [ ] A TOAD TADPOLE WHOSE BACK LEGS ARE JUST STARTING TO GROW?
- [ ] TINY GREEN ALGAE INSIDE SPOTTED SALAMANDER EGGS, PROVIDING FOOD AND OXYGEN?
- [ ] SPOTTED SALAMANDER LARVAE JUST HATCHING, WHICH LOOK LIKE TINY DRAGONS WITH LONG, FEATHERY GILLS?
- [ ] A JEFFERSON SALAMANDER WITH TINY NEW ARMS AND LEGS, SPOTS, AND GILLS THAT LOOK LIKE PUPPY EARS?
- [ ] TINY FAIRY SHRIMP AND MOSQUITO LARVAE—SOME FAVORITE FOODS OF OLDER, GROWING SALAMANDER LARVAE?